ROUTE 66

David Knudson

SHIRE PUBLICATIONS

Published in Great Britain in 2012 by Shire Publications Ltd, Midland House, West Way, Botley, Oxford OX2 0PH, United Kingdom.

44-02 23rd Street, Suite 219, Long Island City, NY 11101, USA.

E-mail: shire@shirebooks.co.uk www.shirebooks.co.uk

A CIP catalogue record for this book is available from the British Library.

Shire Library no. 675 . ISBN-13: 978 0 74781 132 9

David Knudson has asserted his right under the Copyright, Designs and Patents Act, 1988, to be identified as the author of this book.

Designed by Myriam Bell Design, UK and typeset in Perpetua and Gill Sans.
Printed in China through Worldprint Ltd.

12 13 14 15 16 10 9 8 7 6 5 4 3 2 1

COVER IMAGE
Known as "The Mother Road" and "America's Main Street," Route 66 has long captured the hearts and minds of travelers venturing from Chicago to Los Angeles. (By kind permission of Evangelos Souglakos.)

TITLE PAGE IMAGE
Opened in 1938, Roy's was more than a desert mirage for weary travelers.

CONTENTS PAGE IMAGE
It was common for travelers to caravan together for physical and moral support as they drove across the desert. Here a caravan stops, likely to let cars and passengers alike cool down.

PHOTO CREDITS
All images courtesy of the author and the National Historic Route 66 Federation unless otherwise stated.

Shire Publications is supporting the Woodland Trust, the UK's leading woodland conservation charity, by funding the dedication of trees.

CONTENTS

OATS

THE BIRTH OF ROUTE 66

THE BEGINNING

Cyrus Stevens Avery (August 31, 1871–July 2, 1963) of Tulsa, Oklahoma, was known as the "Father of Route 66." Born in Stevensville, Pennsylvania, Avery eventually became one of Oklahoma's best-known highway advocates and civic leaders.

After moving with his parents to Missouri in 1881, young Avery had little formal education. At the age of nineteen, however, he qualified to teach in a country school. He then worked his way through William Jewell College in Liberty, Missouri, graduating in 1897. In 1901, he moved to what was then known as Oklahoma Territory and sold life insurance in Oklahoma City. In 1904, he moved his business to Vinita, Oklahoma, and expanded into real estate loans. Then, foreseeing a strong future in the emerging oil industry, he relocated to Tulsa and launched Avery Oil and Gas Company in 1907.

Avery's interests were varied, reaching far beyond the oil industry to farming, real estate development, and government. He served on the Tulsa

Opposite: From the 1930s through the 1940s, it is estimated that 210,000 people fled the Dust Bowl in the Midwest in all manner of conveyance and migrated to California. John Steinbeck immortalized their plight in his epic novel, *The Grapes of Wrath*.

Left: Goldroad/Sitgreaves Pass, Arizona, is an early stretch of Route 66 between Oatman and Kingman that was modern in its day. The road was challenging to build and remains a challenge to motorists today. The harrowing stretch was mercifully bypassed by the divided, two-lane interstate early in 1953.

Cyrus Stevens Avery of Tulsa, Oklahoma, was known as the "Father of Route 66." He was nearly singlehandedly responsible for the development of Route 66.

County Commission from 1913–16 and on the Tulsa Water Board in the early 1920s, where he was instrumental in bringing water to the rapidly growing city.

Avery believed that well-planned and maintained roads and a system of interstate highways would bring prosperity to Oklahoma and its towns and cities. He became an avid member of several transcontinental road associations that were working to improve roads throughout the country, including the Oklahoma Good Roads Association, the Albert Pike Highway Association, and the pre-World War I National Ozark Trail Association.

Modern roads were not appreciated by everyone, however, particularly in the farm belt. Many opposed the taxes they generated, and those who drove slow-moving farm equipment along the roads often felt threatened by faster vehicles. But Avery saw beyond that fear to the future—a future that would be driven by automobiles.

Avery soon spearheaded the National Ozark Trail Association. (The trail itself eventually evolved into U.S. Highway 66.) He served as vice president of the U.S. 66 Highway Association in 1927, and the U.S. Department of Agriculture placed him on the Joint Board of Interstate Highways from 1925–27. During this time, he also served as one of Oklahoma's highway commissioners from 1922 through 1926.

The 1920s saw the notion of a national highway system take off. Legislation for public highways had first appeared in 1916 with the Federal Aid Road Act, with revisions made in 1921 with the Federal Aid Highway Act, but it was not until Congress amended and authorized existing legislation in 1925 that the government executed its plan for national highway construction.

Opposite, bottom: A look at the Farmersville filling station in Illinois.

Avery was one of the national highway system's biggest supporters. He avidly promoted the concept of a transcontinental highway between Chicago and the Pacific Coast. His dreams were realized when they finally merged with the national program of highway and road development in the 1920s.

A natural salesman, Avery wanted the highway to be designated as "Route 66" because he thought the double sixes were catchy and that the name would make the new road easy to promote for business purposes. But the numbers did not fit into the numerical template established for American highways at the time. Instead, the numbers 60 and 62 were preferred by many because they fit the highway system template. Avery continued to fight for the name, but highway officials and politicians alike battled him for months, nearly defeating his idea—an idea that would become famous.

At the last minute, Avery prevailed and "66" was assigned to the Chicago-to-Los Angeles route in the summer of 1926. Route 66 was commissioned as a federal highway on November 11 of the same year.

From the beginning, public road planners intended U.S. 66 to expand the nation's economy by connecting the main streets of rural and urban communities. At the time, many small towns had no prior access to a major national thoroughfare. There was no way to predict the sweeping changes Route 66 would effect in the years ahead.

PROMOTING ROUTE 66

Almost from the outset, various leaders, promoters, and entrepreneurs were extolling the virtues of the new highway. The first notable Route 66 promotion took place from March 4 through May 26, 1928. It was a footrace called the "Bunion Derby," and it set the pace for the many races, cruises, caravans, rides, and walks that continue to take place today.

The Bunion Derby was the brainstorm of sports promoter and entrepreneur Charles C. "C. C." Pyle (1882–1939), often called "Cash and Carry Pyle." He was a theater owner and

The unlikely winner of the 3,423-mile "Bunion Derby" was Andy Payne, a Cherokee Indian farm boy from Foyil, Oklahoma. The footrace was the first substantial Route 66 promotion and was followed by an endless stream of races, cruises, caravans, rides, and walks that continue to take place along the road today.

The warmer year-round temperatures along Route 66 meant truckers could drive and deliver goods across country many more days than they could on other transcontinental highways that were subject to winter weather difficulties.

sports agent who represented a number of sports celebrities, including football star Red Grange of the Chicago Bears. Grange, known as the "Galloping Ghost," agreed to help Pyle with his transcontinental marathon.

The Bunion Derby was not totally a Route 66 event because it veered off the road, stretching beyond Chicago. It traversed much of 66 but was actually a coast-to-coast footrace beginning in Los Angeles and ending in New York City. Corporate sponsors paid to advertise as the official product or service of the Bunion Derby, and photos, postcards, programs, and other memorabilia were sold along the racecourse. Pyle wasn't the only one who stood to make money from the race; the first-place winner was promised a $25,000 cash prize, a tidy sum at the time (roughly $325,000 in 2012 dollars).

Gathering at the starting line on the first day of the race were 199 world-class runners, as well as scores of amateurs and beginners. Contestants came from all over the world, including the Philippines, Greece, Germany, Estonia, and England. And then there was Andy Payne.

Payne (1907–77) heralded from the Route 66 town of Foyil, Oklahoma, about 40 miles northeast of Tulsa. A Cherokee Indian farm boy, his credentials for the race consisted of being a good runner in high school. While visiting in California, he heard about the race, begged and borrowed the $125 entrance fee, and began training.

Almost from the beginning of the race, injuries took their toll on contestants. They ran across Arizona, New Mexico, and Texas and then into Oklahoma, suffering snow, sleet, rain, and blistering heat. Participants were given a daily food allowance, and many of them dashed into cafés along the racecourse to buy sandwiches—literally on the run.

As the runners entered Payne's home state of Oklahoma, he held a sizeable lead, despite his relative lack of experience. Thousands of onlookers, including Governor Henry Johnston and humorist Will Rogers, gave him a hero's welcome, lining the streets as bands played, cheering him as he ran across the state.

Not everyone had as much endurance as Payne. On May 24, 1928, only fifty-five remaining runners stumbled across the finish line in New York City's Madison Square Garden, where roughly twenty thousand spectators had gathered. Payne was declared the winner. After all was totaled, he had run 3,423 miles in 573 hours, 4 minutes, and 34 seconds, finishing hours ahead of the second-place runner. Claiming his prize money of $25,000, he headed back to Oklahoma, paid off his father's farm, and married his high school sweetheart.

THE FAIR WEATHER HIGHWAY

Promotions and spectacles weren't the only attractions driving people to the new highway. A rapidly changing America sparked demand for Route 66. Although major thoroughfares were not entirely new to the landscape of the United States, this new route was different. Contrasted with the Lincoln, the Dixie, and other major highways of its day, Route 66 did not follow a traditionally linear course. Instead, its diagonal course heading southwest from Chicago linked hundreds of remote, rural communities in Illinois, Kansas, and Missouri, thus enabling farmers to more easily transport grain and produce for redistribution. The new highway's unconventional route cut across several states, making it particularly significant to the trucking industry, which by 1930 had come to rival the railroads for preeminence in American shipping. In addition, this shortened route between Chicago and the Pacific Coast crossed over largely

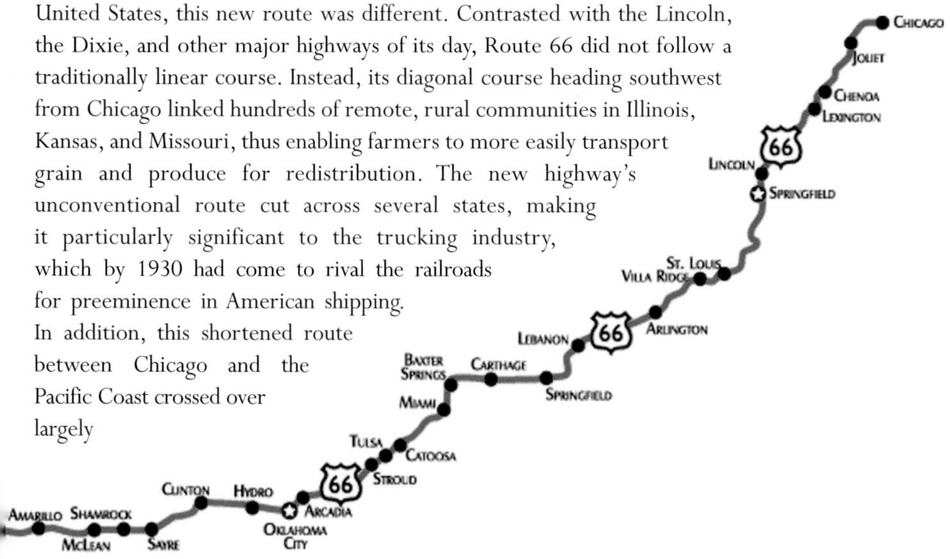

This 1940s photo of Arizona's Painted Desert Inn illustrates that the National Park Service got its share of the income derived from Route 66 tourists by building a nice selection of facilities to attract motorists.

flat prairie lands through a milder climate than in northern highways. This flat, diagonal highway stretching across fairly temperate regions made Route 66 especially appealing to truckers, earning it the title as the "fair weather highway."

But the trucking industry wasn't the only one to benefit from the new highway. The completion of Route 66 on the eve of World War II was particularly beneficial to the military as the new highway could more easily carry troops and convoys across the country from base to base.

The cross-country experience of one young Army captain left an indelible impression. Dwight D. Eisenhower (October 14, 1890–March 28, 1969) found his command bogged down in spring mud near Fort Riley, Kansas, while participating in a coast-to-coast maneuver. It was clear to Eisenhower that the War Department needed more than just muddy, unreliable roads if it was going to move troops and materiel across the country. He called for improved highways in order to speed up mobilization during wartime and to promote national defense during peacetime.

With World War II looming, the War Department looked to the West as the ideal region for military training bases, partly because of its geographic isolation and especially because it offered consistently dry weather for air and field maneuvers. From 1933 to 1938, as part of Roosevelt's "New Deal" economic programs during the Great Depression, thousands of unemployed young men from virtually every state in the country were put to work on road gangs to pave the final stretches of the road. As a result of this Herculean effort, the Chicago-to-Los Angeles highway was reported as "continuously paved" in 1938. Route 66 helped facilitate the single greatest wartime manpower mobilization in the nation's history.

The Boots Motel opened in 1939 in Carthage, Missouri. It is one of the few remaining examples of Streamline Moderne architecture, which was once very popular along Route 66.

ROADSIDE ARCHITECTURE

People working and living along Route 66 recognized early on that even the most frugal of travelers had to eat and sleep and had to fill up the gas tank. Tourist courts, garages, and diners soon popped up all along the highway. If military use of the highway during wartime ensured the early success of roadside businesses, the demands of the new tourism industry in the postwar decades gave rise to modern facilities that guaranteed long-term prosperity.

The evolution of tourist-targeted facilities is well represented in the roadside architecture along U.S. Highway 66. For example, most travelers who drove the route did not stay in hotels. Instead, they preferred the easy convenience of the accommodations that emerged from this golden age of automobile travel: auto camps, motor courts, and motels.

Soon Route 66 eating establishments joined the Streamline Moderne trend. This architectural style lasted through the 1950s and encompassed everything from automobiles to toasters.

Outside Newburg, Missouri, John's Modern Cabins opened not long after the commissioning of Route 66. Cabins were the transition between early tent camps and motels. Ongoing efforts to preserve what is left of this property have not made much headway.

This is a beautifully restored example of a Valentine Diner on display at the Route 66 Museum in Clinton, Oklahoma. They were prefabricated as eight-to-ten-seat diners that one or two people could operate.

From the 1920s through the 1950s, many unusual examples of Roadside Vernacular architecture sprung up along Route 66. Gradually, building codes spelled the end of this fanciful construction. The La Cita Restaurant in Tucumcari, New Mexico, leaves little doubt that it serves Mexican food.

Wigwams were common examples of Roadside Vernacular on Route 66. Here, Tee Pee Curios in Tucumcari, New Mexico, sells Indian souvenirs.

In Sanders, Arizona, the Sanders Route 66 Diner's architecture can best be described as eccentric. However, it serves the same purpose as Roadside Vernacular: it demands attention.

Auto camps developed as townspeople along Route 66 roped off spaces in which travelers could camp, usually in tents, for the night. Camp supervisors—some of whom were employed by the governments of the various states through which the highway passed—provided clean water, wood for fires, toilets and showers, and even laundry facilities, often free of charge. As auto camps became more sophisticated, they morphed into motels, which many travelers preferred.

The national outgrowth of the auto camp and tourist homes was the cabin camp (sometimes called cottages), which offered informal, minimal comfort at affordable prices. As early as the late 1920s and early 1930s, Amarillo, Texas, boasted more than twenty-five of these tourist courts; a few of these cottages are still in operation in various states. These cottages were much more substantial than the original auto camps. Many of the buildings were permanent, winterized structures, and some even featured attached garages.

Eventually, auto camps gave way to motor courts, where all of the guest quarters were housed under a single roof. Motor courts and motels offered additional amenities, such as adjoining restaurants, souvenir shops, and swimming pools. Among the more famous still in operation along Route 66 are the Munger Moss Motel in Lebanon, Missouri, which originally was opened as a barbecue place and, in 1946, expanded into a motel, and the El Rey Inn in Santa Fe, New Mexico, which opened in 1936 with just twelve rooms.

Motels grew in favor because they were considered the most modern lodgings along America's roadsides. They were designed and built in many different styles, ranging from elaborate and flamboyant to spare and dreary. Seasoned Route 66 travelers claimed they could determine the price of a room simply by the style of a motel.

The Aztec Hotel in Monrovia, California, on an early alignment of Route 66 was designed by noted architect Robert Stacy-Judd. Its span of nearly a half a block of Mayan-themed architecture is impossible to miss.

Filling stations and motels are the most popular restoration projects along Route 66. Although a few of the stations have been restored to their original use, environmental regulations have made this a costly proposition. Therefore, repurposing is most common. This is the 1932 Odell Standard Station, which was beautifully restored by the Illinois Route 66 Association and now serves as a gift store.

Many people believe the first motel was built on Route 66, but Interstate Highway 101 actually claims that title. Los Angeles architect Arthur Heineman actually built the world's first motel in 1925 in San Luis Obispo, California, about halfway between San Francisco and Los Angeles. Heineman coined the term "motel," meaning motor hotel, and called his new venture the "Milestone Mo-Tel," which it certainly became (the motel closed in 1991).

Lodging was by far not the only industry to spring up along the new highway. Like the first motel, many believe that the first filling station appeared on Route 66. In fact, however, the first facility in the world to provide petrol for customers is considered by historians to be a pharmacy in Weslock, Germany, which began the service in 1888. Although debatable, it is generally believed that the first purpose-built filling station in the United States opened in 1907 in St. Louis, Missouri, but not on Route 66.

Even so, filling stations evolved readily along Route 66. In the early years of the new highway, station prototypes were developed regionally by petroleum companies. Those that were considered successful were adopted and repeated across the country. Most stations featured a simple one-story frame house with one or two service pumps out front. As they became more successful, they became more elaborate with the addition of service bays and tire outlets. Among the more interesting examples of the evolution of gas stations that are

Cool Springs Station, outside of Oatman, Arizona, has performed as a movie set and is now repurposed as a very nice gift shop.

The movie version of *The Grapes of Wrath* was a box-office success and received critical acclaim as well. John Ford won an Academy Award for Best Director.

still along Route 66 are the Standard Station in Odell, Illinois; the Tower/U-Drop Inn complex in Shamrock, Texas; and the DX Station in Afton, Oklahoma.

ROUTE 66 IN POPULAR MEDIA

The completion of Route 66 in 1938 couldn't have come at a better time for the thousands who traversed it while fleeing the Plains for greener pastures. From the 1930s through the 1940s, an estimated 210,000 people migrated from the East and Midwest to California to escape the despair of the Dust Bowl. Enormous, wicked clouds of black dirt and brown dust swept across the prairie lands of the Plains, making it nearly impossible to breathe and turning farms into barren fields. The choking dust was everywhere—inside homes, schools, and shops, covering cars, barns, and streets. For those who endured that particularly painful experience, and in the view of generations of children to whom they recounted their stories, Route 66 symbolized the "road to opportunity."

John Steinbeck memorialized the plight of Dust Bowl farmers in his controversial novel *The Grapes of Wrath*, which became the landmark Route 66 book and film. In his Pulitzer- and Nobel prize-winning social commentary, Steinbeck (1902–68) proclaimed U.S. Highway 66 "The Mother Road." His epic 1939 novel, combined with John Ford's 1940 film adaptation of the grueling odyssey of the Joad family fleeing the Dust Bowl along Route 66, served to immortalize the highway in the American consciousness.

Born in Salinas, California, Steinbeck spent most of his life in California's Monterey County, the setting of much of his fiction. Though he attended Stanford University intermittently between 1920 and 1926, he never graduated, instead supporting himself through manual labor while pursuing a career as a writer. Steinbeck's first three novels were, for the most part, critical and commercial failures. His first real success came with the 1935 publication of *Tortilla Flat*, an affectionate story about Mexican Americans. With the publication in 1936 of his novella *Of Mice and Men*, Steinbeck achieved some acclaim and notoriety. Though he would go on to write more than twenty-five novels, it was *The Grapes of Wrath* that would remain his crowning literary achievement.

Of course, Steinbeck wasn't the only one to feature Route 66 in his creative work. In fact, since the highway's origins in the 1930s, it has been featured in countless books, magazines, and films. For example, although

George Maharis, left, and Martin Milner played Buz Murdock and Tod Stiles respectively in the ground-breaking TV series *Route 66*. It was the first series filmed almost entirely outdoors rather than on sound stages. It has been said that the Corvette was the third star of the show. Considered very realistic in the 1960s, today's audiences will find it rather quaint.

Bobby Troup's song "(Get Your Kicks on) Route 66" did every bit as much to embed the Route 66 name in the minds of the public as did *The Grapes of Wrath* and the 1960s TV series. It originally was released by the Nat "King" Cole Trio in 1946, and Cole reported that throughout his career, he received more requests for it than any other song he had recorded.

never mentioned specifically in the book, many believe that Sal Paradise likely traveled at least partially along the Mother Road in Jack Kerouac's 1957 classic *On the Road*.

Route 66 and many points of interest along the way were familiar landmarks by the time a new generation of postwar motorists hit the road in the 1960s. It was during this period that the television series, *Route 66*, starring Martin Milner as Tod Stiles and George Maharis as Buz Murdock, drove their Corvette into the living rooms of America every Friday evening. The show ran on CBS from October 7, 1960, to March 13, 1964. By today's standards, it is rather naïve, but in the 1960s, it was realistic action and it brought many Americans out to the route looking for new adventure.

The series made history by being the first filmed almost entirely on location. Milner reports, however, that only one episode was actually filmed on or even near 66. It was shot in Chicago. The producer thought other parts of the country, particularly the Northwest, afforded more interesting scenery than the real route.

In addition to literature and film, the Mother Road also made its way into song. After World War II, thousands of Americans, including many former soldiers, sailors, airmen, and Marines, left the harsh winters of the East and Midwest for the balmier climes of the West and Southwest. Many hopped on Route 66, driving across the country to their new hometowns. One such emigrant was Robert William Troup, Jr., of Harrisburg, Pennsylvania. Bobby Troup (1918–99), an ex-Marine captain, penned a lyrical roadmap of the now-famous cross-country road. The lyrics from the 1946 song "(Get Your Kicks on) Route 66" became a catch phrase for countless motorists who traveled from Chicago to the Pacific Coast with stops in St. Louis, Missouri; Amarillo, Texas; and Gallup, New Mexico, among other cities in the eight states through which Route 66 passes. First recorded by the Nat "King" Cole Trio, it became an instant classic, which has since been covered by dozens of artists, from Chuck Berry (1961) to The Rolling Stones (1964). The English band Depeche Mode combined the tune with its own composition in 1987's "Behind the Wheel." In 2006, the song was featured in Pixar's animated movie *Cars* and in the movie *RV* with Robin Williams.

ROUTE 66 BECOMES OUTDATED

Although Route 66 endures in popular culture, its legend long outlived its usefulness. Excessive truck use during World War II and the resurgence of the automobile industry immediately following the war put great pressure on America's highways. Much of the national highway system had become nearly undriveable during and soon after the war years. Just about every highway had become functionally obsolete and even dangerous because of narrow pavements and antiquated supports with reduced carrying capacity.

By the mid-1950s, public demand for rapid mobility and a better highway system portended the demise of Route 66. President Dwight Eisenhower, who had returned from Germany impressed by Hitler's autobahn, enjoyed public support and federal sponsorship for his interstate system of divided multilane highways. "During World War II," he said, "I saw the superlative system of German national highways crossing that country and offering the possibility, often lacking in the United States, to drive with speed and safety at the same time." Congress agreed, and the passage of the Federal Aid Highway Act of 1956 provided a comprehensive financial umbrella to

After World War II, Route 66 had deteriorated severely, rendering stretches of it virtually undriveable. This 16-foot-wide section of road west of Nilwood, Illinois, was dangerously narrow for increasingly larger vehicles, and the pavement had degraded to the point where it became suited only for agricultural traffic.

As remnants of Route 66 were bypassed, some stretches were deeded back to the original landowners. This section near the now-defunct town of Landergin, Texas, can no longer be driven or explored because it is on fenced-off private property.

underwrite the cost of the national interstate and defense highway system, spelling the beginning of the end for Route 66. In less than two decades, nearly all segments of the Mother Road were bypassed or paved over by new stretches of modern four-lane highways.

Despite the demise of the highway, Route 66 had paved the way for the evolution of highway development in the United States. In just a few decades, the nation's corridors had evolved from a rudimentary hodge-podge of state and county roads to a federally subsidized complex of uniform, well-designed interstate expressways. Various alignments of the Route, most of which can still be found, illustrate the evolution of road engineering from coexistence with the surrounding landscape to domination of it.

From its origins in the interwar years to its popularity among postwar travelers, Route 66 moved generations of families across the country. After the economic catastrophe of the Great Depression, after the devastation of the Dust Bowl, and after the horrors of global war, it symbolized the renewed spirit of optimism that pervaded the country. Often called "The Main Street of America," it linked a remote and underpopulated region with two vital twentieth-century cities: Chicago and Los Angeles.

Though the Main Street of America endures in the hearts and minds of many, it could not withstand the pace of progress. U.S. Highway 66 bowed to the multilane interstate highway system on October 13, 1984, when the final section of the original road was bypassed by Interstate 40 at Williams, Arizona. It ceased to be a federal highway when it was decommissioned on June 25, 1985.

The contribution this legendary road made to the nation must be evaluated in the broader context of American social and cultural history. Its appearance on the American scene coincided with unparalleled economic strife and global instability, yet it hastened the most comprehensive westward movement and economic growth in U.S. history. Like the early trails of the nineteenth century, Route 66 helped to spirit a second but perhaps more permanent mass relocation of Americans. Although most of those trails have disappeared, Route 66 has experienced a renaissance in recent years, brought to pass by many dedicated people, organizations, and local, state, and federal government groups.

This stretch of Route 66 east of Bellemont, Arizona, deteriorated, then was bypassed, and finally closed altogether.

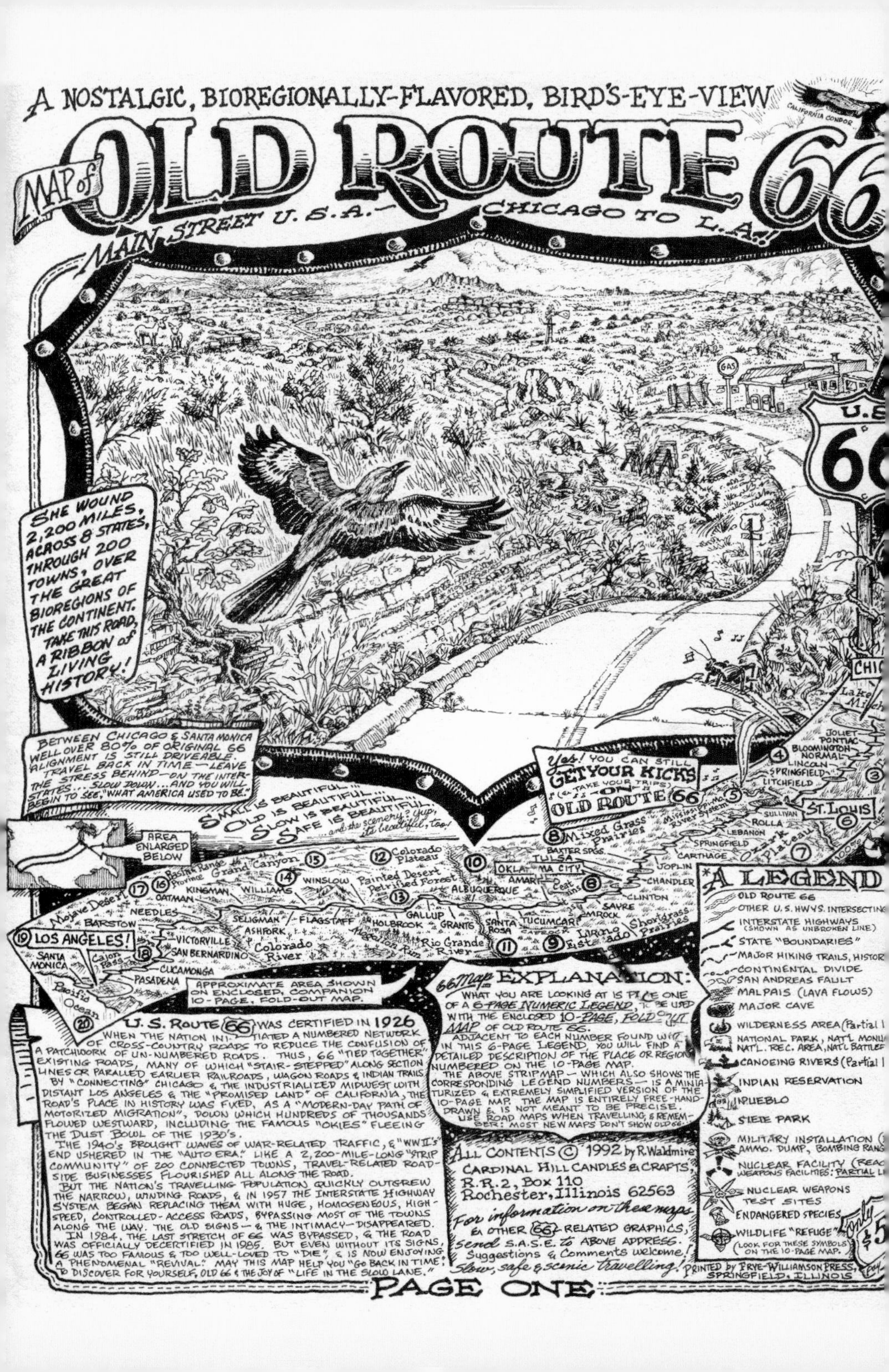
A NOSTALGIC, BIOREGIONALLY-FLAVORED, BIRD'S-EYE-VIEW
CALIFORNIA CONDOR
MAP of OLD ROUTE 66
MAIN STREET U.S.A. – CHICAGO TO L.A.!
GAS
SHE WOUND 2,200 MILES, ACROSS 8 STATES, THROUGH 200 TOWNS, OVER THE GREAT BIOREGIONS OF THE CONTINENT. TAKE THIS ROAD, A RIBBON of LIVING HISTORY!
BETWEEN CHICAGO & SANTA MONICA WELL OVER 80% of ORIGINAL 66 ALIGNMENT IS STILL DRIVEABLE. TRAVEL BACK IN TIME — LEAVE THE STRESS BEHIND — ON THE INTERSTATES... SLOW DOWN... AND YOU WILL BEGIN TO SEE, "WHAT AMERICA USED TO BE."
SMALL IS BEAUTIFUL... OLD IS BEAUTIFUL... SLOW IS BEAUTIFUL... SAFE IS BEAUTIFUL ...and the scenery? yup, it's beautiful, too!
Yes! YOU CAN STILL GET YOUR KICKS (& TAKE YOUR TRIPS) ON OLD ROUTE 66
AREA ENLARGED BELOW
JOLIET
PONTIAC
BLOOMINGTON
NORMAL
LINCOLN
SPRINGFIELD
LITCHFIELD
ST. LOUIS
SULLIVAN
ROLLA
LEBANON
SPRINGFIELD
CARTHAGE
Ozark Plateau
JOPLIN
BAXTER SPGS.
TULSA
OKLAHOMA CITY
CHANDLER
CLINTON
SAYRE
AMARILLO
TUCUMCARI
SANTA ROSA
Mixed Grass Prairies
Mississippi River System
Shortgrass Prairies
Llano Estacado
Rio Grande River
ALBUQUERQUE
GALLUP
GRANTS
Painted Desert Petrified Forest
Colorado Plateau
HOLBROOK
WINSLOW
FLAGSTAFF
WILLIAMS
ASHFORK
SELIGMAN
KINGMAN
OATMAN
Grand Canyon
Basin & Range Province
Colorado River
NEEDLES
Mojave Desert
BARSTOW
VICTORVILLE
SAN BERNARDINO
Cajon Pass
CUCAMONGA
PASADENA
LOS ANGELES!
SANTA MONICA
Pacific Ocean
APPROXIMATE AREA SHOWN ON ENCLOSED, COMPANION 10-PAGE, FOLD-OUT MAP.
U.S. ROUTE 66 WAS CERTIFIED IN 1926 WHEN THE NATION INITIATED A NUMBERED NETWORK OF CROSS-COUNTRY ROADS TO REDUCE THE CONFUSION OF A PATCHWORK OF UN-NUMBERED ROADS. THUS, 66 "TIED TOGETHER" EXISTING ROADS, MANY OF WHICH "STAIR-STEPPED" ALONG SECTION LINES OR PARALLED EARLIER RAILROADS, WAGON ROADS & INDIAN TRAILS.
BY "CONNECTING" CHICAGO & THE INDUSTRIALIZED MIDWEST WITH DISTANT LOS ANGELES & THE "PROMISED LAND" OF CALIFORNIA, THE ROAD'S PLACE IN HISTORY WAS FIXED, AS A "MODERN-DAY PATH OF MOTORIZED MIGRATION", DOWN WHICH HUNDREDS OF THOUSANDS FLOWED WESTWARD, INCLUDING THE FAMOUS "OKIES" FLEEING THE DUST BOWL OF THE 1930'S.
THE 1940'S BROUGHT WAVES OF WAR-RELATED TRAFFIC, & "WWII'S" END USHERED IN THE "AUTO ERA." LIKE A 2,200-MILE-LONG "STRIP COMMUNITY" OF 200 CONNECTED TOWNS, TRAVEL-RELATED ROAD-SIDE BUSINESSES FLOURISHED ALL ALONG THE ROAD.
BUT THE NATION'S TRAVELLING POPULATION QUICKLY OUTGREW THE NARROW, WINDING ROADS, & IN 1957 THE INTERSTATE HIGHWAY SYSTEM BEGAN REPLACING THEM WITH HUGE, HOMOGENEOUS, HIGH-SPEED, CONTROLLED-ACCESS ROADS, BYPASSING MOST OF THE TOWNS ALONG THE WAY. THE OLD SIGNS — & THE INTIMACY — DISAPPEARED.
IN 1984, THE LAST STRETCH OF 66 WAS BYPASSED, & THE ROAD WAS OFFICIALLY DECERTIFIED IN 1985. BUT EVEN WITHOUT ITS SIGNS, 66 WAS TOO FAMOUS & TOO WELL-LOVED TO "DIE", & IS NOW ENJOYING A PHENOMENAL "REVIVAL." MAY THIS MAP HELP YOU "GO BACK IN TIME," TO DISCOVER FOR YOURSELF, OLD 66 & THE JOY OF "LIFE IN THE SLOW LANE."
66 Map EXPLANATION:
WHAT YOU ARE LOOKING AT IS PAGE ONE OF A 6-PAGE NUMERIC LEGEND, TO BE USED WITH THE ENCLOSED 10-PAGE, FOLD-OUT MAP OF OLD ROUTE 66.
ADJACENT TO EACH NUMBER FOUND WITHIN THIS 6-PAGE LEGEND, YOU WILL FIND A DETAILED DESCRIPTION OF THE PLACE OR REGION NUMBERED ON THE 10-PAGE MAP.
THE ABOVE STRIPMAP — WHICH ALSO SHOWS THE CORRESPONDING LEGEND NUMBERS — IS A MINIATURIZED & EXTREMELY SIMPLIFIED VERSION OF THE 10-PAGE MAP. THE MAP IS ENTIRELY FREE-HAND-DRAWN & IS NOT MEANT TO BE PRECISE.
USE ROAD MAPS WHEN TRAVELLING, & REMEMBER: MOST NEW MAPS DON'T SHOW OLD 66.
ALL CONTENTS © 1992 by R. Waldmire
CARDINAL HILL CANDLES & CRAFTS, R.R.2, Box 110 Rochester, Illinois 62563
For information on these maps & OTHER 66 RELATED GRAPHICS, send S.A.S.E. to ABOVE ADDRESS.
Suggestions & Comments welcome!
Slow, safe & scenic travelling!
A LEGEND
OLD ROUTE 66
OTHER U.S. HWYS. INTERSECTING
INTERSTATE HIGHWAYS (SHOWN AS UNBROKEN LINE)
STATE "BOUNDARIES"
MAJOR HIKING TRAILS, HISTOR
CONTINENTAL DIVIDE
SAN ANDREAS FAULT
MALPAIS (LAVA FLOWS)
MAJOR CAVE
WILDERNESS AREA (Partial
NATIONAL PARK, NAT'L MONU
NAT'L. REC. AREA, NAT'L BATTLEF
CANOEING RIVERS (Partial
INDIAN RESERVATION
PUEBLO
STATE PARK
MILITARY INSTALLATION
AMMO. DUMP, BOMBING RANG
NUCLEAR FACILITY (REAC
WEAPONS FACILITIES: PARTIAL
NUCLEAR WEAPONS TEST SITES
ENDANGERED SPECIES
WILDLIFE "REFUGE"
(LOOK FOR THESE SYMBOLS ON THE 10-PAGE MAP.)
Only $5
PRINTED by FRYE-WILLIAMSON PRESS, SPRINGFIELD, ILLINOIS
PAGE ONE

REVITALIZING THE MOTHER ROAD

AFTER ROUTE 66 WAS BYPASSED by the interstate highway system and then decommissioned in 1985, it lapsed into anonymity, its heyday mostly forgotten. Businesses that thrived on the thousands of cars and trucks that drove the highway each day closed down as traffic all but disappeared. America was in a hurry, so drivers didn't want to waste time on a slow two-lane road when they could race along on the multilane turnpikes at 70 miles per hour. Travelers could drive hundreds of miles very quickly, though seeing very little and stopping only for fuel and a quick meal. This suited most people just fine because travelers could now get from point A to point B in half the time. Drivers bragged about and compared their driving times with others. Major tourist destinations thrived.

AVERTING DEATH

But not everyone was enamored with the high-speed interstates. Angel Delgadillo of Seligman, Arizona, understood the mystique and magic of the old road—something no one ever claimed for the turnpikes.

Opposite: Bob Waldmire produced dozens of illustrations of Route 66. Most were intricately detailed, bird's-eye views.

Left: Angel Delgadillo, surrounded by walls of memorabilia in his Seligman, Arizona, barber shop, is credited with being the first to recognize the historical importance of Route 66, then raising the awareness of its value to America.

Delgadillo was born on April 19, 1927, in a house on the main thoroughfare in Seligman, Arizona, situated along Route 66. The dirt road out front was the main east–west artery and would become the paved Route 66 within a few years after his birth.

Delgadillo and his eight siblings, along with his parents Angel and Juanita Delgadillo, grew up watching the traffic flow by on 66. During the Dust Bowl years, when thousands migrated west from the dust-strewn Plains, they saw people driving by in their automobiles loaded with all of their worldly possessions. "It was an amazing caravan of poor folks heading west, seeking opportunities to better their lives," Delgadillo recalled.

The Depression and Dust Bowl years were tough times for the Delgadillo family, too—so tough that they also considered loading up the family Model T to follow the migration west. But instead they hung on in Seligman. In order to make a little money and keep the family together, Delgadillo and his siblings formed a musical group and played Big Band tunes at towns and cities all along Route 66.

Seligman, Arizona, is one of many Route 66 communities to have capitalized on the road's notoriety by catering to the many tourists from around the world who explore it. The "Rusty Bolt" is a startling sight that is a store that carries everything from biker apparel to Route 66 memorabilia and is festooned with mannequins.

Delgadillo soon followed in his father's footsteps, attending the American Pacific Barber College in Pasadena, California, also along Route 66. He served as an apprentice from 1948 to 1950 in Williams, Arizona, another Route 66 town. He began shaving beards and clipping hair in Seligman in 1950 and continued through 1996 when he retired. During those years, Delgadillo wed his sweetheart and raised four children in Seligman.

In September 1978, Angel Delgadillo watched his hometown wither when the Interstate Highway system replaced Route 66 as the primary means for automobile travel through Seligman. For Delgadillo and many others who worked and raised their families along the Route, it was a sad day, a day when they realized that the world had just about literally bypassed them. Like many of the towns along Route 66, Seligman had been condemned by the very brand of progress that originally energized it: a new, faster highway system. Soon businesses closed, residents left, travelers stopped coming, and buildings fell into disrepair.

Although some residents and town leaders looked for ways to attract new industry to the dying town, Delgadillo had a different idea. He believed that the past held the key to the future of Seligman, and that the history, memory, and myth of Route 66 would pave the way for the survival of his town.

In February 1987, Delgadillo and others founded the Historic Route 66 Association of Arizona. By November of that year, they had successfully lobbied the Arizona Legislature to dedicate U.S. Route 66, from Seligman to Kingman, as "Historic Route 66," a designation that later was bestowed upon all of Arizona's Route 66. Soon after, the seven other states (California, New Mexico, Texas, Oklahoma, Kansas, Missouri, and Illinois) along Route 66 also formed associations. Six years later, the National Historic Route 66 Federation was founded.

The Mother Road has charmed many artists into trying to capture her unique personality. One of the first to fall under her spell was Bob Waldmire from Springfield, Illinois. His unique illustrations and free-spirit lifestyle made him likely the most famous Route 66 personality.

ILLUSTRATING THE MOTHER ROAD

Around the time that Angel Delgadillo was launching the Historic Route 66 Association of Arizona, Bob Waldmire was tacking small, paper "Route 66" signs on poles and fences along the Route. (Most of the original Route 66 shield signs were gone, removed by the Highway Administration or collectors or simply vandalized beyond use.) Waldmire, in his inimitable low-key manner, hoped his little signs might aid travelers in finding and following the old road.

Waldmire was born on April 19, 1945, in St. Louis, Missouri. He and his family soon moved to Springfield, Illinois, where they started the Cozy Dog Drive-in on Route 66. Waldmire's father Edwin is said to have invented the "hot dog on a stick" concept, yet fans of the drive-in—as well as most Route 66 enthusiasts—will argue that Cozy Dogs are far superior to common corn dogs.

Although corn dogs might not have been Bob Waldmire's professional calling, Route 66 had struck a chord in his heart, and it manifested itself through art. "I always liked to draw, especially small things. I started drawing them as a kid," he told writer Bob Venners in *Desert Exposure*, "and I've never stopped."

Waldmire attended Southern Illinois University in Carbondale, Illinois, about 170 miles south of Springfield. While at college, Waldmire saw a fellow student's "bird's-eye-view" rendition of Carbondale, and it struck a nerve. When he returned home, he drew a picture of his hometown. Though it was, by his own admission, an amateur effort, local merchants liked the idea and soon were paying him to include their businesses in a poster version of the drawing.

At that moment, Waldmire realized he could do what he loved—draw—for a living. With that, Waldmire eventually created bird's-eye-view posters of thirty-four cities, many of them college towns. He worked closely with local merchants to help them promote their businesses, which sold more posters—a profitable win–win partnership. His success made him realize that he could do this work just about anywhere. He could head to warmer climes in the winter, and this he did—using Route 66 as his guide, drawing all along the way.

Waldmire captured the icons of Route 66: motels (the Wigwam Motel in Rialto, California), restaurants (Steve's Café in Chenoa, Illinois), and gas stations (Soulsby's Shell Station in Mt. Olive, Illinois)—even weird places (the Edsel graveyard in Shamrock, Texas), whole towns (Needles, California), and stretches of the old road (in Hydro, Oklahoma; Halltown, Missouri; and Dwight, Illinois).

Soon his drawings of Route 66 and its icons were printed on postcards and posters. Within a matter of time, people began requesting custom-drawn imagery of the Route. By 1985, Waldmire's love for Route 66 was even more pronounced, and he realized that he needed a vehicle he could live out of while he drew and peddled his art. He soon purchased an orange, 1972 Volkswagen Microbus. The VW became his primary residence for nearly twenty-five years—and a Route 66 icon itself. It was the inspiration for the animated character of "Fillmore" from the Pixar movie *Cars*.

Waldmire garnered substantial accolades over the years. None, however, did he cherish more than the 2004 John Steinbeck Award bestowed upon him just five years before his death for his efforts to preserve, promote, and restore the legendary highway.

MEMORIALIZING THE ROUTE

Although numerous authors have produced innumerable books recounting the many characteristics of Route 66, two of the earliest were also the most influential: Michael Wallis's *Route 66: The Mother Road* (1992; reissued in 2001 as a seventy-fifth anniversary edition) and Tom Teague's *Searching for 66* (1991).

Michael Wallis was born in 1945 just off Route 66 in St. Louis. After attending a string of public and private schools, putting in a hitch with

Michael Wallis wrote the first comprehensive book regarding Route 66, *Route 66: The Mother Road*. He has authored sixteen other works, and his voice can be heard as the sheriff in Pixar's animated *Cars* movies.

the U.S. Marines, and bouncing around a few colleges and universities, he opted for life as a writer, working day jobs as a ranch hand, bartender, hotel waiter, social worker, printer, and ski-lodge manager. Citing such literary and artistic greats as Ernest Hemingway, Thornton Wilder, Jack Potter, and Dorothy Brett as his mentors, he managed to finally earn his wings as a writer. Still, he never forgot his roots or those formative years, and to this day he remains a staunch child of the sixties.

Wallis has received the John Steinbeck Award, the Arrell Gibson Lifetime Achievement Award from the Oklahoma Center for the Book, the Will Rogers Spirit Award, and the Western Heritage Award from the National Cowboy Hall & Western Heritage Museum. He is a member of the Oklahoma Writers Hall of Fame, Writers Hall of Fame of America, and the Oklahoma Historians Hall of Fame, and was the first inductee into the Oklahoma Route 66 Hall of Fame. He has been nominated for the Pulitzer Prize three times.

Tom Teague (1943–2004) was born in Kansas City, Missouri. He self-published his book *Searching for 66* in 1996 with limited distribution, which accounted for less visibility than Wallis's *Route 66: The Mother Road*. Unlike Wallis's large-format book, which includes numerous photos and chronicles the history of 66, Teague's publication is a collection of stories about people he met during his trips along the road. It features the intricate artwork of Bob Waldmire.

Small in stature, Teague dreamed big—and he delivered on those dreams. A U.S. Army veteran who taught high school and sold fishing bait, one of his passions was Route 66. In fact, he founded and served as the first president of the Route 66 Association of Illinois, and he created the Route 66 Hall of Fame at the Dixie Trucker's Home in McLean, Illinois. He led a campaign to post five hundred Route 66 signs in Illinois, making it easier to chart and follow the original Route 66 to this day.

Teague was also the founder of the Soulsby Station Society. The Soulsby Station in Mt. Olive, Illinois, was originally built as a Shell Station in 1926 by

Henry Soulsby (1910–99). The building ranks as one of the oldest filling stations still standing on Route 66. Teague led the preservation effort that saved this treasure and helped it become named to the National Register of Historic Places. The National Historic Route 66 Federation presented him with the John Steinbeck Award in Amarillo, Texas, at the 1999 International Route 66 Festival.

Today, many Route 66 icons share a place on the National Register, including Lou Mitchell's Restaurant in Chicago; Red Cedar Inn in Pacific, Missouri; Provine Service Station in Hydro, Oklahoma; the Ranchotel in Amarillo, Texas; the KiMo Theater in Albuquerque, New Mexico; the Peach Springs Trading Post in Peach Springs, Arizona; and the Harvey House Railroad Depot in Barstow, California. Many of these sites are open to visitors; the National Park Service features a list of Route 66 sites at its website (www.nps.gov/history/nr/travel/route66).

THE NATIONAL HISTORIC ROUTE 66 FEDERATION

The National Historic Route 66 Federation can be justly credited with expanding Route 66 beyond the efforts of a few. This was accomplished by promoting the highway around the world, establishing a global Route 66 organization, and producing events along the route that featured authors, artists, musicians, and collectors. In addition, the Federation's John Steinbeck Awards honored those working to restore the highway.

The roots of the Federation go back to October 1964, when David Knudson drove from Chicago to California on Route 66. Fresh out of college, he had pocket change and plenty of dreams—but no job. He landed in California and eventually built a business in Los Angeles, but he never forgot his trip along Route 66 with all the fancy motor courts, exotic trading posts, and intoxicating aromas of sweet smoke from the pit barbecues. He couldn't afford to stop at any of them on his way out, but he vowed that one day he would travel across Route 66 again and buy some Indian moccasins, sample the great-smelling barbecue, and stay in a few places with clean sheets.

Tom Teague wrote *Searching for 66* and founded the Route 66 Association of Illinois, the Illinois Route 66 Hall of Fame, and the Soulsby Station Society.

Lucille's Provine Station was built and opened for business in 1929 but gained fame when it was bought by the Hamons family in 1941. Lucille Hamons was so outgoing, friendly, and helpful to travelers that she gained the nickname "Mother of the Mother Road."

Thirty years later, in August 1994, his chance came when Knudson and his wife Mary Lou decided to drive from Chicago to California along Route 66. But they couldn't find it. The famous road wasn't on any maps, and there were no "66" road signs. The old road had been bypassed by an interstate highway, stranding many of the once-thriving businesses and towns. Deserted structures stood as silent reminders of the days of America's Glory Road.

By the time David and Mary Lou arrived home, they decided to sell their business interests and devote their time to trying to save as much as possible of the historic road before it was completely gone. The Federation was born.

Today, the National Historic Route 66 Federation is a worldwide, nonprofit organization whose mission is:

> "directing the public's attention to the importance of U.S. Highway Route 66 in America's cultural heritage and acquiring the federal, state and private support necessary to preserve the historic landmarks and revitalize the

economies of communities along the entire 2,400-mile stretch of road. The Federation accomplishes these goals through public education, advocacy and membership activities. Public outreach strategies include publication of the quarterly magazine *Federation News*, the Adopt-A-Hundred Preservation Program, a worldwide web site www.national66.org, serving on the National Route 66 Corridor Preservation Program's Grant Committee, the *Route 66 Dining and Lodging Guide*, the *EZ66 GUIDE for Travelers*, and assistance to the media, authors, learning institutions and production companies."

David and Mary Lou Knudson founded the National Historic Route 66 Federation after driving Route 66 in 1994. Shortly thereafter, they helped write and spearhead the passage of the National Route 66 Preservation Bill, which issues grants to Route 66 property owners.

In 1999, the National Route 66 Preservation Bill was passed by Congress and signed into law by President Bill Clinton. The act provided $10 million in matching fund grants to individuals, corporations, and communities for the purpose of preserving or restoring historic properties along the legendary route. The Federation spearheaded this bill for more than four years and is working with the National Route 66 Corridor Preservation Program to issue the grants.

NATIONAL ROUTE 66 CORRIDOR PRESERVATION PROGRAM

In 1999, in response to the National Route 66 Preservation Bill, the Route 66 Corridor Preservation Program was created. Administered by the National Park Service, the program collaborates with private property owners; nonprofit organizations; and local, state, federal, and tribal governments to identify, prioritize, and address Route 66 preservation needs. It provides cost-share grants for the preservation and restoration of the most significant and representative properties dating from the route's period of outstanding historical significance (1926–70). These properties include businesses related to gas, food, and lodging; cultural landscapes; and the all-important road segments themselves. Cost-share grants also are provided for research, planning, oral history, interpretation, and education and outreach projects related to Route 66.

The program serves as a clearinghouse of preservation information and provides limited technical assistance. It is administered by the National Park Service's National Trails System Office in Santa Fe, New Mexico.

Hill Top
MOTEL
BEST VIEW IN KINGMAN
APPROVED
AAA
MOTEL
REFRIGERATOR - MICROWAVE
REMOTE TV 6 - HBO
FAX - INTERNET
LAUNDROMAT

EXPLORING ROUTE 66

Even today, the National Historic Route 66 Federation receives letters and e-mails from people around the world asking if they risk being attacked by Indians in the West. This misconception seems to stem from early western TV shows and movies that are still playing in many countries depicting Indians as wanton savages—a gross overstatement even in the heyday of the cowboy-and-Indian feuds. Suffice to say that travelers are not in danger from Native Americans, although they may lose some money in one of their many gambling casinos along the route.

The surest way to enjoy a trip along Route 66 is to do your homework and plan in advance. Strangely, a bewildering number of people blithely fly to Chicago, rent a car, and start looking for Route 66. But the route is not on ordinary maps and it is poorly signed. This is not to discourage those who prefer some serendipity in their traveling—people who like to roam as the moment strikes them rather than having a firm schedule. Regardless of your preferences, you will want, at the very least, the *EZ66 GUIDE for Travelers* by Jerry McClanahan. Without it, serendipity likely will turn into frustration.

Opposite: In recent years, many historic Route 66 properties have been lovingly restored. The Hill Top Motel is one of many in Kingman, Arizona, a popular destination community for Route 66 travelers.

Left: As the architecture of Route 66 businesses reached out for the tourist dollars, so did road signs for businesses. The signs lined the roadside, some repeating for many miles, growing more urgent as a motorist drew closer to the establishment. After hundreds of miles of signs for the Jack Rabbit Trading Post west of Joseph City, Arizona, this was the last.

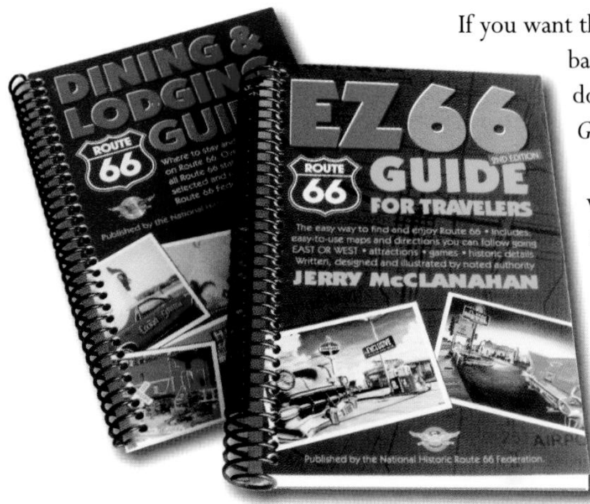

If you want the freedom to "do your own thing," have back-up information available while you are doing it: *The Route 66 Dining & Lodging Guide* is a good source.

For instance, let's say you are traveling west and want to stay outside Los Angeles and explore the city the next day. It's late when you arrive in San Bernardino, and you discover there isn't a room available from there to Los Angeles because the Auto Club Speedway in Fontana is holding its annual National Hot Rod Association drag races. So, you can get out the guide and call a motel in Victorville to hold a room.

Route 66 is not on ordinary maps and there are not many road signs. So, it is important to get the right materials and do some preplanning before starting out on the legendary road.

Such spur-of-the-moment planning may not be ideal, but it can be the price of freedom. This kind of freedom may be liberating, yet it also means that you may miss opportunities to stay at some of the more unique and popular places because they are booked, particularly during the busier months.

One particularly alarming story was reported by the National Historic Route 66 Federation a few years ago. Many auto enthusiasts from around the world fly or ship their classic cars over to tour Route 66. A group of Swedish buffs flew their 1950s American cars over, gathered in Chicago at a

The Painted Desert Indian Center still beckons tourists to watch for them and turn in to browse through their Indian souvenirs.

Geronimo, another Joseph City, Arizona, Trading Post, uses their signs to promote "wondrous" sights.

hotel, and left early the next morning to travel Route 66. Before they left, they asked the hotel manager where Route 66 was, so he directed them to it. Incredibly, all they really knew about the famous road was that it went from Chicago to Santa Monica. They found it but became immediately lost, so they stopped and asked at a gas station who called the Federation to get directions for them because there were no "Route 66" road signs for them to follow. Over the next week, they would call periodically to get directions. Several times their calls came from many miles from the route.

The sign "Mountain Lions" advertises a long-gone mountain lion pit in Arizona. These once-common wildlife attractions do not exist anymore because they would be judged inhumane by today's standards.

Meramec Caverns, southeast of Stanton, Missouri, also has signs along Route 66 for hundreds of miles in both directions, the major difference being that they were all painted on barns. This avoided the costly process of continually replacing stolen and damaged roadside signs.

Certain times of year are better than others to travel the route. Two factors are in play here: weather and geography. No matter what time of year you travel, you will not run into heavy traffic or throngs of people, except possibly in several of the major cities. If you have the choice, you will find

either spring or autumn the most comfortable seasons along Route 66. Summer can get very hot in the Mojave Desert. Winter can produce considerable snow and ice in the Midwest and the higher elevations of Arizona. However, many people do drive the route during these times because that is when they get their vacation time. If winter and summer are the only choices available to you, be prepared. Make sure to drive a well-maintained vehicle with operable heat and air conditioning. Allow a little extra time in midwinter because it is very possible you will get snowed in for a day or two in the mountains near Flagstaff.

People travel the route in and on many different types of vehicles, but we recommend that you don't drive a recreational vehicle for one good reason: You will miss much of the essence of the legendary Mother Road. Route 66 was and is a commercial entity that was constructed to carry travelers across the country and developed to provide services and entertainment for them on their trips.

Today, vintage motels, diners, stores, and attractions have come to symbolize the route to people around the globe. Fortunately, many of these icons of a bygone era remain today and have been restored to their original condition. In keeping with the early days, their owners offer old-fashioned hospitality and their prices are usually considerably lower than chain

Opposite, bottom: This sign sculpture on Route 66 for Red Oak II advertises the curious village northeast of Carthage, Missouri, which was accumulated, restored, and built by noted artist Lowell Davis.

Below: This Route 66 "sign" advertises nothing. The leaning water tower is outside Groom, Texas, and the name "Britten" is simply the last name of the wealthy rancher who owns the land it is on.

operations off the interstate highways. If you drive a recreational vehicle, you will have little or no opportunity to, for example, sleep in a wigwam or enjoy a pony shoe sandwich.

If driving a recreational vehicle and camping along Route 66 is, however, your dream vacation, you will want to explore the Route 66 RV Campground listings at http://rwarn17588.wordpress.com/campgrounds and http://www.facebook.com/US66Travel or get one of the Woodall RV Guides or Trailer Life RV Guides and cross-check it with a Route 66 guide or map in order to find campgrounds.

Other travelers opt for more organized trips along the route. Bus tours of Route 66 can be problematic, however. There are quite a few available, most arranged by brokers in countries other than the United States. Unfortunately, many of these tours tend to spend most of their time racing down interstates, only occasionally skimming the actual route. Because most bus tour passengers are from non-English-speaking countries and have only a cursory understanding of Route 66, it is likely many such travelers rarely see the original road, let alone some of the most interesting and scenic stretches. This is certainly a deceptive practice, but organizers often argue they have to keep up a brisk pace to be profitable. An additional problem the tour operators face are the many smaller villages along the route that do not have facilities to accommodate a sudden influx of approximately sixty bus passengers.

If you have a choice, spring and autumn are the best times of year to travel Route 66. Although today's air-conditioned vehicles are built to withstand extreme heat, the desert is still not the best place to be in the summer months.

A few years ago, the Smithsonian Institution and a major tour company ran several excellent Route 66 bus tours that featured noted authors and historians. However, the cost of dining and lodging for the span of the tours made them very expensive, so they were discontinued only after a few trips. The situation is unfortunate because a well-prepared bus tour is appealing to many. Someone else does the driving and planning and takes care of reservations and luggage. If you are considering a bus tour, get personal recommendations from passengers who have taken a tour that interests you.

There are several high-quality and very successful guided driving and riding tours of Route 66 conducted by entrepreneurs and organizations from an assortment of countries. Most of them are motorcycle tours that require you to bring your own motorcycle or rent one.

The most popular mode of traveling the route is a personal vehicle. A car, truck, or motorcycle can provide unmatched freedom and flexibility. You can stop and go where and when you want. But is it better to drive your own vehicle or rent one? If you are considering driving your own, you will want to consider wear and tear on it, and you will need the time to drive both directions.

Renting is very popular but can get expensive. Not so much because of the daily rental and mileage rates but the drop-off fee. If you return the vehicle to a different location than where you began, chances are good you will be charged a hefty drop-off fee. However, this fee can be avoided if you return the vehicle to its original location. The best thing to do regarding rental cars is to go to the websites of the major international car rental agencies (Alamo, Avis, Budget, Dollar, Hertz, National, Thrifty, etc.) or try www.carrentals.com and compare. Rates are very erratic and change constantly. Contrary to popular belief, with the exception of small local agencies, which you will want to avoid, no single agency is consistently cheaper than another. It all depends on when you will be traveling, your departure point and destination, how long you will be gone, the type of vehicle you would like, how much they want your business at the moment, and so forth. Because of all the Route 66 travelers renting cars and driving east to west (the customary direction), the agencies have many more cars in the West. So, when you make comparisons, you will find the drop-off fee for your car will likely be less if you drive west to east.

The National Historic Route 66 Federation regularly gets requests for information about renting classic cars for Route 66 trips. Many of these inquiries come from people who want to relive the life that Buz and Todd once did in the TV series *Route 66* and rent an early Corvette. Unfortunately, no one rents classic cars anymore for long-distance travel. The insurance and maintenance costs are simply too high.

PLACES TO VISIT

Like any tourist destination, Route 66 has scores of "must see" places. These are some of the highlights along the 2,400 miles of the Route. The places start in Chicago and move west.

Lou Mitchell's restaurant in Chicago is a great place to start a tour of Route 66.

Lou Mitchell's Restaurant. 565 West Jackson Boulevard, Chicago. Telephone: 312-939-3111. Website: www.loumitchellsrestaurant.com

If you are starting your Route 66 trip in Chicago, you can't go wrong starting it with breakfast at Lou Mitchell's, as have millions since the route was commissioned. It opens at 5:30 a.m. on weekdays so you can get an early start. This institution is the quintessential big-city diner with a flair—they give you free donut holes and Milk Duds—a flair they don't really need because their food is outstanding.

Illinois and Missouri are home to two Route 66 towns with striking murals.

Pontiac, Illinois, Murals. Telephone: 815-844-5847. Website: http://visitpontiac.org

Many Route 66 travelers favor the farmland and small, historic villages of Illinois that still have so much of the early route intact. Two Route 66 communities boast dozens of striking murals within their borders: Cuba, Missouri, and Pontiac, Illinois. Expect to take the better part of a day in Pontiac as you browse the murals and shops and the well-done Route 66 Museum. Remarkably, eighteen of these beautiful murals were created in four days by a group of mural artists called the "Walldogs." While there, try one of the town's vintage cafés.

Cozy Dog Drive In. 2935 S. Sixth Street, Springfield, Illinois. Telephone: 217-525-1992. Website: www.cozydogdrivein.com
Although they look pretty much alike, you'll immediately taste the difference between a Cozy Dog and an ordinary corn dog. Dipped in batter and deep-fried, the Cozy's batter jacket has a vivid crunch and the dog within is plumper and juicier. The business goes back to Ed Waldmire, Bob Waldmire's father, in 1946. The Waldmire family still owns and operates it. No serious Router would ever pass it by.

A cozy dog is no ordinary corn dog, and the Cozy Dog Drive In is the place to go for the best.

Ariston Café. 413 North Old Route 66, Litchfield, Illinois. Telephone: 217-324-2023. Website: www.ariston-cafe.com
Begun in 1924 by Pete Adam, his son, Nick Adam, and his wife, Demi, continue the family tradition of serving what many Routers consider the best dinners anywhere on the route. You'll also appreciate the classic Route 66 interior, which includes a vintage neon sign with this understatement, "Remember—Where Good Food Is Served."

Henry's Rabbit Ranch. 1107 Historic Old Route 66, Staunton, Illinois. Telephone: 618-635-5655. Website: www.henrysroute66.com
The visitors' center is in a replica of an early Route 66 filling station, and that's where the fun begins. Congenial Rich Henry's place is worth the stop, and the tons of lovable bunnies and the emporium of highway memorabilia

The Ariston Café has been a Route 66 classic since 1924.

Henry's Rabbit Ranch offers a plethora of memorable souvenirs.

are bonuses. It's difficult to guess how many of these cute little guys sold themselves to Routers and moved on down the road.

Old Chain of Rocks Bridge. St. Louis, Missouri, and Madison, Illinois. Website: www.theroadwanderer.net/66Illinois/chain.htm

This unusual bridge carried Route 66 traffic across the Mississippi from Illinois to Missouri from 1929 through 1967. Its most conspicuous feature is a 22-degree bend in the middle which narrows the bridge, creating an accident-prone junction. This peculiarity was necessary to establish sound

The Old Chain of Rocks Bridge carried traffic across the Mississippi River from 1929 until 1967. The 22-degree bend in the bridge made it accident-prone.

The original Ted Drewes Frozen Custard opened in 1931.

footings in the treacherous channel. The bridge was saved from demolition and restored and is now open to pedestrian traffic only.

Ted Drewes Frozen Custard. 6726 Chippewa, St. Louis, Missouri. Telephone: 314-481-2652. Website: www.teddrewes.com/home/default.aspx
Ted Drewes has two locations in St. Louis. This one, on Route 66, the original, was opened in 1931 by Ted Drewes, Sr., and is run today by his son, Ted Drewes, Jr. It is best known for "concretes"—a custard blended with your choice of dozens of ingredients and served in a large yellow cup with a spoon and straw. The straw is all but useless because the treat is so thick that the spoon won't even fall out if the cup is turned upside down.

Munger Moss Motel. 1336 East Route 66, Lebanon, Missouri. Telephone: 417-532-3111. Website: www.mungermoss.com
This circa 1946 Route 66 motel has several good reasons to stop by. The owners are two of the best ambassadors the route has. They greet you like family and are encyclopedias of Mother Road information. The place is beautifully and faithfully restored, and the rooms are spotless. While in Lebanon, visit the exceptional Lebanon-Laclede Route 66 Museum.

This 1940s motel is classic Route 66 lodging.

Afton Station. Afton, Oklahoma. Telephone: 918-382-9465. Website: http://postcardsfromtheroad.net/afton.shtml
This is Laurel Kane's labor of love. It is easy to simply drive by this place and admire the impressively restored filling station. But don't. Pull up and go in. The interior is every bit as impressive as the exterior. Kane is one of the route's

Laurel Kane's Afton Station features a collection of classic Packards.

foremost authorities and a very cordial hostess. She might even show you the classic Packard collection. But to make sure she's open, call ahead first.

POPS. 660 West Highway 66, Arcadia, Oklahoma. Telephone: 405-928-7677. Website: http://route66.com/25.0.html
This is the newest landmark on Route 66. Its can't-miss design features a 66-foot soda bottle that glows at night in an infinite array of colors and patterns. Inside the futuristic building you'll find a record-breaking selection of nearly six hundred ice-cold sodas and beverages. Many of the brands you'd long forgotten, like Dr. Brown's Cel-Ray Soda, Faygo Orange, and Frostie Root Beer. Pops is already known for their exceptional burgers and breakfasts.

POPS is one of the route's newest landmarks.

The recently restored Blue Whale provides plenty of fun and relaxation for weary travelers.

Blue Whale. 2705 North Highway 66, Catoosa, Oklahoma. Telephone: n/a. Website: www.roadsideamerica.com/tip/1017
Privately owned entertainment centers proliferated along Route 66 in the early days. After hot and dusty hours on the road, they provided fun and relaxation for children and adults alike. Usually, these attractions included picnic areas and a refreshment stand. Just such a place is the recently restored Blue Whale in Catoosa, Oklahoma, which also includes a swimming pond and an in-whale water slide.

Oklahoma Route 66 Museum. 2229 West Gary Boulevard, Clinton, Oklahoma. Telephone: 580-323-7866. Website: www.route66.org/index2.html
There are quite a few very well done Route 66 museums along the road, but one stands out: the Oklahoma Museum. Opened in 1995, little expense was spared.

The Oklahoma Route 66 Museum is one of the best museums along the route.

Visitors take a narrated "trip" down the Mother Road through dozens of creative exhibits that include the history of transportation, lodging, dining, garages, and curio shops. Don't miss the beautifully restored Valentine diner on the museum grounds.

U-Drop Inn. 101 East Twelfth Street, Shamrock, Texas. Website: www.nps.gov/nr/travel/route66/tower_station_u-drop-inn_cafe_shamrock.html
The most unusual building on a road known for the unusual is this one-time restaurant/filling station built in 1936. It was designed by John Nunn, the owner of the land, who scratched the design in dirt with a nail and showed it to a contractor. Saved from the wrecking ball, it was purchased in 1999 by a local bank and gifted to the town. It currently houses their Chamber of Commerce. Look for it in the animated movie *Cars*.

This unusual building was built in 1936 and was featured in the movie *Cars*.

Big Texan Steak Ranch. Just off the Route at 7701 East Interstate 40, Amarillo, Texas. Telephone: 800-657-7177. Website: www.bigtexan.com
Eat an entire 72-ounce steak here and they'll give it to you free! They are really 72-ounce roasts. This is a very colorful and rowdy place you don't want to miss. If you're not game for the 72-ouncer, get a smaller one because they are quite good. While in the area, stop by the Cadillac Ranch with the old graffittied Cadillacs buried nose down in the ground west of Amarillo.

The Big Texan is home of the 72-ounce steak.

Cadillac Ranch has become one of the defining images of Route 66.

Cadillac Ranch. Just west of Amarillo, Texas, on Route 66. Telephone: n/a. Website: www.roadsideamerica.com/story/2220
Eccentric millionaire Stanley Marsh 3 (he insists on "3" rather than "III"), for reasons known only to him, hired the Ant Farm art collective from San Francisco to bury ten Cadillacs nose down in one of his fields. He won't talk about it, so myths and rumors abound. One thing is absolutely true, however: The Cadillac Ranch is the defining image of the Mother Road. Travelers are tacitly encouraged to walk through the field's unlocked gates and spray-paint whatever they want on the cars.

MidPoint Café. Adrian, Texas. Telephone: 806-538-6379. Website: http://uglycrustpies.com
In the mid-1990s, Fran Houser, owner of the famed MidPoint Café, decided that Adrian, Texas was midway between Chicago and Los Angeles. Then she opened her very popular MidPoint Café. Some in surrounding communities debate the midway status, but nobody argues with her if they want to sample some of the best comfort food on Route 66. No matter what you order, it is wonderful, but her true claim to fame is her "ugly crust pie." The atmosphere is pure 1950s 66, and the gift shop is one of the best along the route.

Located halfway between Chicago and Los Angeles, the MidPoint Café offers some of the best comfort food on the Mother Road.

Blue Swallow Motel. 815 E. Route 66 Boulevard, Tucumcari, New Mexico. Telephone: 575-461-9849. Website: www.blueswallowmotel.com
There is no lodging place on Route 66 more

Beautifully restored, the Blue Swallow Motel is one of the most storied buildings on the route.

storied than this. But why? Is it because it is the quintessential Route 66 motel? Is it because it has stuck it out in a town barely hanging on? Is it because it is restored right down to the telephones? Or because it was owned for years by a famous lady? What about that famous neon? Yes to all of the above. Keep in mind that the motel is closed October through February.

Santa Fe Route 66. Website: www.theroadwanderer.net/66NMex/santafe.htm
An early alignment of Route 66 in New Mexico went up 60 miles from Albuquerque to Santa Fe. The Route followed the old El Camino Real/Santa Fe Trail north to the ancient, renowned La Fonda Hotel. As interesting as Old Town Santa Fe is, though, you will find a wealth of vintage Route 66 motels that still line Cerrillos Road. You can almost feel the enchantment of the early tourists as they settled in for the night and dreamed of old Mexico.

A number of vintage motels like the El Rey still line parts of Route 66.

The El Rancho Hotel was built in 1937 and housed plenty of movie stars.

El Rancho Hotel. 1000 E. Highway 66, Gallup, New Mexico. Telephone: 800-543-6351. Website: www.elranchohotel.com
Gallup and the scenery around it were once prime locations for western movies, not to mention the ready availability of Indian extras. This hotel was built in 1937 for the sole purpose of housing stars and crews as they cranked out movies, sometimes several a day. Many of its rooms are named after stars who stayed in them, including Ronald Reagan and John Wayne. The awesome lobby alone is worth the stop.

Petrified Forest National Park. 1 Park Road, Petrified Forest, Arizona, 26 miles east of Holbrook, Arizona. Telephone: 928-524-6228. Website: www.nps.gov/pefo/contacts.htm
This National Park Service attraction is famous for its petrified trees, which lived about 225 million years ago. It has been mistakenly rumored that the petrified material is no longer there, pilfered by souvenir hunters and enterprising business owners. In fact, there are plenty of these amazing artifacts to see, thanks mostly to a law that makes it illegal to remove the material. In addition to the fossils, there is an extensive museum and gift store as well as a dinosaur park.

Petrified Forest National Park is home to plenty of amazing artifacts.

Hackberry General Store. 11255 East Highway 66, Hackberry, Arizona. Telephone: 928-769-2605. Website: http://hackberrygeneralstore.com
In 1992, Route 66 artist, environmentalist, and all-around icon Bob Waldmire opened this rustic

Route 66 icon Bob Waldmire opened this funky store in Arizona.

store in scenic Crozier Canyon, which was made famous on Santa Fe Railway posters. As it became a very popular stop, there is no question that it also contributed to Waldmire's notoriety. Then he announced he was weary of all the traffic and being awakened by tourists in the middle of the night, so he sold it to John and Kerry Pritchard. The Pritchards soon put their own personality into the unique place. With a 1956 Corvette in front of the store functioning as a seat for a blonde mannequin, it is impossible to miss. The interior is pure 1950s eclectic that challenges visitors to see everything in a day.

La Posada Hotel. 303 E. Second Street, Winslow, Arizona. Telephone: 928-289-4366. Website: www.laposada.org
Many consider this to be the finest place to stay along the entire route. It is beautiful inside and out, it's quiet despite the trains going by now and then, it has a legendary history, and the food is wonderful. Originally, it was an elegant Harvey House bustling with prim and proper Harvey Girls (Harvey House waitresses). The Santa Fe railroad owned the Harvey Houses and the tracks that brought travelers to their doors to stay or simply to dine. Few were ever disappointed.

La Posada Hotel is considered one of the finest along Route 66.

Seligman, Arizona. Website: www.theroad wanderer.net/RT66seligman.htm
Founded in 1895, this town has retained its Route 66 flavor better than most others. Nearly all of the buildings that were there during the heyday remain and few have been added since. Around 250 hearty souls live here today.

Visitors enjoy the Snow Cap, Angel's barber/gift shop, Seligman Sundries, the Roadkill Café, the Copper Cart Restaurant, the Rusty Bolt, the Supai Motel, and the Route 66 Motel.

The Snow Cap Drive In is just one of the many wonderful places in Seligman, Arizona.

Powerhouse Visitors Center. 120 W. Andy Devine Avenue, Kingman, Arizona. Telephone: 928-753-9889. Website: www.kingmantourism.org/Welcome_to_Kingman
The visitors center is housed in the town's historic powerhouse building. It is located in the heart of the longest remaining stretch of Route 66. The center's Route 66 museum depicts the evolution of travel along the Mother Road, displaying murals, photos, and life-size dioramas. It also features one of the best gift shops of Route 66 items anywhere.

Oatman, Arizona & Sitgreaves Pass. Website: www.oatmangoldroad.org
It is safe to say that there is no town even a little like Oatman, anywhere in the world. In 1915, the discovery of gold quickly turned this place into a mining town of 3,500 people—then it went bust. Prospectors fled and left their burros to fare for themselves, which the ancestors of those critters do quite well now, thanks to handouts from tourists. Going down the mountain east of town is heart-pounding Sitgreaves Pass. If you are not adverse to mountain roads without guardrails, this is a scenic winner.

The historic Powerhouse building is home to Kingman's visitors center.

There are few places along the route like Oatman, a former mining town.

Roy's Café and Gas. 87520 National Trails Highway, Amboy, California. Telephone: 760-733-1066. Website: www.rt66roys.com
Roy's is Amboy. The town was founded by salt miners in 1858, but little was there until Buster Burris opened Roy's in 1938. It was undoubtedly a welcome sight to thousands of early Route 66 travelers—in the middle of the Mojave Desert—to find gas, a place to spend the night, get some food, and wait for auto repairs. Deserted for many years, restaurateur Albert Okura bought the entire town in 2005 and is gradually restoring it.

Route 66 Mother Road Museum. 681 N. First Avenue, Barstow, California. Telephone: 760-255-1890. Website: www.route66museum.org
This unique Route 66 museum is in the nicely restored Casa del Desierto Harvey House. It emphasizes the history of the Route 66 corridor through the Mojave Desert area. It includes many displays and photographs unique to the region. In addition, it features revolving exhibits from Mother Road artists and photographers.

The Route 66 Mother Road Museum is housed in the restored Casa del Desierto Harvey House.

Summit Inn. 5970 Mariposa Road, Hesperia, California. Telephone: 760-949-8688. Website: www.jeffreysward.com/tributes/summitic.htm
This is the consummate roadhouse. It is everything Route 66 travelers envision when they think of Route 66: great comfort food, cheery yet efficient waitresses, and walls covered with 1950s memorabilia. The

inn, in fact, sits on the summit of the awesome Cajon Pass offering breathtaking views of the San Bernardino and San Gabriel Mountains. In the early days of Route 66, the pass was the liberating end to the threatening Mojave Desert and the welcome beginning to enchanting Southern California.

The Summit Inn offers breathtaking views and great food.

California Route 66 Museum. 16825 South D Street, Victorville, California. Telephone: 760-951-0436. Website: http://califrt66museum.org
This museum is packed with a variety of exhibits and artifacts ranging from the very sincere to the very quirky. In the quirky category is one of the quirkiest Route 66 icons—Hula Ville—a traffic-stopping collection of large wooden and metal cutouts crafted by Miles Mahan, the biggest being a 12-foot-tall hula girl.

Wigwam Motel. 2728 W. Foothill Boulevard, Rialto, California. Telephone: 909-875-3005. Website: www.wigwammotel.com
This Route 66 icon opened in 1949 to shouts from a generation of kids: "I want to stay in a teepee!" Seven of these fanciful places were built around the country; two remain on Route 66. This one has undergone extensive restoration by the owners and is now back to near original. The grounds are beautifully maintained including a swimming pool and numerous palms.

Barney's Beanery. 8447 Santa Monica Boulevard, West Hollywood, California. Telephone: 323-654-2287. Website: http://barneysbeanery.com
This is the original, which opened in 1920. There are five sites now, but the others don't successfully replicate this wacko place. Virtually every celebrity

The California Route 66 Museum features a collection of quirky and iconic exhibits and artifacts.

This Route 66 icon, the Wigwam Motel, opened in 1949.

has at one time or another been seen here. In fact, it is almost guaranteed you'll see at least one star when you stop in. In the early days, notables would "camp" here all day and night sampling the spirits and atmosphere. Apparently, a few still do.

Santa Monica Pier, Santa Monica, California. Website: http://www.santamonicapier.org/ Route 66 enthusiast Dan Rice worked hard to get the pier recognized as the "emotional" end of the Mother Road. He was able to get a sign erected declaring it. There are actually two official ends nearby. Rice owns a Route 66 souvenir shop on the pier, "66-to-Cali," so his efforts were a little self-serving; yet he is a major Route 66 supporter. What better way to end a 2,400-mile trip than to gaze out onto the blue Pacific from atop a Ferris Wheel.

ROUTE 66 MUSEUMS

Barstow Route 66 Mother Road Museum. 681 N. First Avenue, Barstow, California. Telephone: 760-255-1890. Website: www.route66museum.org

California Route 66 Museum. 16825 S. D Street, Victorville, California. Telephone: 760-951-0436. Website: www.califrt66museum.org

Lebanon-Laclede County Route 66 Museum. 915 S. Jefferson, Lebanon, Missouri. Telephone: 417-532-2148. Website: www.lebanon-laclede.lib.mo.us/Museum.history.html/

The original Barney's Beanery, pictured here, opened in 1920.

National Route 66 Museum. U.S. 66 and Pioneer Road, Elk City, Oklahoma. Telephone: 580-225-6266. Website: www.elkcity.com/Pages.asp?s=mus&id=7

Oklahoma Route 66 Museum. 2229 W. Gary Boulevard, Clinton, Oklahoma. Telephone: 580-323-7866. Website: www.route66.org/index2.html

Powerhouse Route 66 Museum. 120 W. Andy Devine Avenue, Kingman, Arizona. Telephone: 928-753-9889. Website: www.kingmantourism.org/to-do-and-see/museums/route-66-museum/index.php

Route 66 Hall of Fame & Museum. 110 W. Howard Street, Pontiac, Illinois. Telephone: 815-844-4566. Website: http://il66assoc.org/attraction/route-66-association-hall-fame-museum

The Santa Monica Pier, at the end of Route 66, features a Ferris Wheel with an amazing view of the Pacific Ocean.

FURTHER READING

Knowles, Drew. *Route 66 Adventure Handbook: Turbocharged Fourth Edition.* Santa Monica Press, 2011.

McClanahan, Jerry. *EZ66 GUIDE for Travelers.* National Historic Route 66 Federation, 2011.

McClanahan, Jerry, Ross, Jim, and Graham, Shellee. *Route 66 Sightings.* Ghost Town Press, 2011.

National Historic Route 66 Federation. *Route 66 Dining & Lodging Guide 15th edition.* National Historic Route 66 Federation, 2011.

Olsen, Russell A. *The Complete Route 66 Lost & Found.* Voyageur Press, 2008.

Wallis, Michael. *Route 66: The Mother Road: 75th Anniversary Edition.* St. Martin's Griffin, 2001.

Wickline, David. *Images of 66* (Volumes 1 and 2). Roadhouse 66, 2006.

INDEX

Numbers in italics refer to illustrations